Magpie Mind

"My magpie mind collects pieces in a puzzle."
p 10, *Bye Bye Blackbird*

Eileen Berry

Magpie Mind
poems of people, place, and change

Eileen Berry

Plain View Press, LLC
1101 W 34th Street, STE 404

www.plainviewpress.net
Austin, TX 78705

Copyright © 2018 Eileen Berry Family Trust. All rights reserved under International and Pan-American Copyright Conventions. No part of this book may be reproduced or distributed in any form or by any means, or stored in a data base or retrieval system, without written permission from the author. All rights, including electronic, are reserved by the author and publisher.

ISBN: 978-1-63210-027-6
Library of Congress Control Number: 2017933620

Acknowledgements

Grateful acknowledgement is made to the following journals in which the named poems first appeared:
The Licking River Review for "The Red Poem"
Tar River Poetry for "Singular"
The Meridian Anthology of Contemporary Poetry for "Chance"
Soundings East for "Eclogue"
Confluence for "Red Straw Hat"
Passager for "Ms. Susan Lives Alone"
Gulf Stream Magazine for "Ars Poetica"
Crossborder Literary Journal for "Isolation"

The poem "Waiting for Something Terrible to Happen" was published in a different form as "The Doll in Riversedge."

Contents

Acknowledgements	4

People   11

Red Straw Hat	13
Ms Susan Lives Alone	14
A Smashing Good Time	15
The Feelings Are Whole	17
A Modern Woman	19
Some Women	20
Dignity	21
Understanding	22
Only Then	23
The Autocrat	24
The Immigrant	25
At the Boot Inn—1956	26
Just to Say (Warm Version)	27
Never	28
Economy	29
This Cold Morning	30
Waiting or Walking or On a Crowded Bus	31
The Shrimpers	32
Family	33

Places   37

Not on the Map	39
St. Cuthberts' Church	41
Ancient Right of Way	42
Only the Wind There Once	43
Helvellyn	44
Before	45
Larch Street and the Recent Past	46
English Summer	47

Happened	48
Ecology	49
Eclogue	50
Last Train	51
Word Hoard	52
Orange Sun Sets on Ottoman Empire	53
Street Scenes	54

Change — 55

Isolation	57
Transience	58
Wildness	59
October Comes Round Again	60
Waiting for Something Terrible to Happen	61
Chance	62
Singular	63
Reflections	64
Reality	65
Drama	66
Hiatus	67
Persistence	68
The Red Poem	69
Rain Dance	70
Absence	71
Separation	72
Ars Poetica	73

Notes	75
About the Author	77

for my dear husband Len

and my precious children

*so much depends
upon*

*a red wheel
barrow
glazed with rain
water*

*beside the white
chickens*

William Carlos Williams

People

Eileen and Len's wedding, Oct 8 1966

Red Straw Hat

Black skeletal bones, thin shins, small
knobbly kneecaps and ankles.
The dark head rests skull-like on a pillow,
coarse grey-white tufts of hair sprouting
here and there, like sheep's wool caught on a fence.
Nose, the most prominent feature, sharp
and essential.
Her skin has the texture of worn leather.
She is happy, euphoric even (from pills),
as blankets slide off.
I want to cover her up.
The daughter tries, but she shoves her away,
In some French patois, she tells us she is happy,
says leave me, I don't need you.
The daughter, plump and jolly, rakishly adjusts
her red straw hat, laughing at herself in the glass.
She happy, she say.
She don't need me no more.
She gurgles with laughter, fruity, juicy
as pineapples.
Maman don't need me no more.

Ms Susan Lives Alone

Ms Susan sits on the porch, watching
and thinks of a time when
nothing stayed put, when the house
was always a mess, had to be tidied up,
when she wanted order, quiet, time for herself,
time to think.
Now, everything stays put, nothing
moves unless she moves it.
Time is extravagant, clocks irrelevant.
She likes to be outside, where there is always
movement and change, squirrels
jumping and shaking tree branches, birds
wheeling and shrieking,
where leaves fall and have to be swept,
and there are dead flower-heads to snip,
beds to weed.
She thinks she ought to get a dog, but fears
it would be for the wrong reason,
like marrying a man one didn't love.

A Smashing Good Time

There was this huge sign in gold lettering above the front door
SEAVIEW PRIVATE HOTEL

Private hotel—in the parlance of the time and place was seen as something above a boarding house but not as grand as a full service hotel and, as it stood on a side street off the promenade, the seaview was limited.

A few deck chairs of the old-fashioned kind, slings of striped canvas on a wooden frame, were placed on the lawn in a small front garden. Here the guests, always referred to as such, would sit—red-faced where they had caught the sun. The women in new summer frocks with white shoes and handbags; the men with trousers rolled-up, knotted white handkerchiefs on their heads.

They were on their holidays, coming to the seaside every summer, by bus or train from industrial towns and cities, to get away from the smoke and grit of the factories. Some came in Wakes Week when whole towns were almost closed down.

What they sought was sea and sun and a nice clean place to stay. They came to get brown, to do nothing.

I didn't have to lift a finger, someone would boast, I'm a lady of leisure.

While they drank their morning coffee—Nescafe with hot milk, or their afternoon tea—hot and strong, poured from a small brown pot—and enjoyed cakes and sandwiches—thin, brown bread and cucumber filling—they wrote postcards to friends at home. "Having a good time—weather glorious" or "Smashing place—lovely weather."

Seaside towns catered to them with variety shows on the pier where the brass bands went tiddley pom-tiddley pom and there would be comic turns and sing-songs.

At places like the SEAVIEW PRIVATE HOTEL they would gather in the early evenings in the formal dining room with its white tablecloths and serviettes, polite silver cruets of salt, pepper and vinegar, and sit down to dinner—to plates of ham and chips, egg, sausage and chips, followed by puddings and pies with yellow custard.

Some would opt instead for the fish and chip shop on the front, carrying out food wrapped in newspaper soaked in vinegar and grease from the fryer. On the way back they might stop at the pub for a pint or two. No beer was served at the private hotel and wine not even thought about. A nice cup of tea was always on offer.

When it came to alcohol the men would confess they liked a drink or two while the ladies would tell each other how much they preferred a nice cup of tea.

When they took the huge gold sign down—way in the future—it was as if a whole world was being swept away but that world of working class holidays at the seaside had already gone by then.

The Feelings Are Whole

*Feelings have no shape or size, obey no
allometric principle*

The place that I remember is an overgrown
patch of grass and weeds at the bottom of an old garden,
beyond the huge sycamore tree.
It is very warm, the sun is high, early afternoon,
one day about 1932. I have been sent out to play,
after the midday meal and I am making a secret house
for myself in the long coarse grass. I am perhaps four
or five years old.
The feelings are of contentment, warmth, of being happily
alone, out-of-doors under an encompassing blue sky and a
shimmering, buzzing, stirring, moving, rustling of grass,
insects and butterflies; myself a part of it.

Memory adds something. My father coming to say goodbye as
he goes back to work. Bicycle clips in hand, he stands
above me. He is joking with me. He calls me "Lana" and "Old
Sport". My name is Eileen. He ruffles my hair and tells me
not to get my dress dirty, I'm not sure of all these details
but it was the way he spoke to me, the way it would have
been. He is part of the day I remember.

There is a photograph in an old album of a little girl
standing in the tall grass at the bottom of the garden in
the old house. My hair is cut short, the sun is in my eyes.
My dress is one made by my mother—a lovely white linen
with embroidery of white on white. It had a fine orange thread
drawn through the stitches round the neck. This is what I
remember but the photograph is black and white.

The photograph is how I see myself in the garden on
that warm summer afternoon but it may not have been so.
In fact, probably not. The dress was one of my best so
it is unlikely I would have been sent out to play in it.
Also if I was playing alone and my father was coming to
check that I was all right before he went to work, he would
not have had me stand there while he took a photograph.
I do not remember having the photograph taken.

The truth is I have no visual memory of myself in that
garden that day at all, only the feeling of warmth and
sun, blue high skies and grass and insects and the
thousands of tiny noises they make together. This feeling
is so whole and so complete that I can shut my eyes now
or look up into the high blue sky of another day and be
back there, sitting down inside the tall thick grass.
Me, myself, alone; the same me.

The visual memory is of a younger me—the little girl
in the photograph, my mother's child.
The visual memory I piece together as I write.

The feelings are whole and it is only in writing that I
break them into pieces.

A Modern Woman

Housework she hated—the boring, endless, repetitive
routine that trapped her in the traditional role of
housewife.
Whenever she could she would be out-of-doors, walking
on the marsh or in the garden.
She loved gardening, grew peas and beans; hollyhocks,
gladioli and roses.
Being outside meant freedom.

Sometimes while she weeded a flower bed or pegged
washing on a clothesline, she would hear the unusual
sound of an aeroplane in the sky, would stop, look up
and see a small silver speck cutting through the high
blue, leaving white trails.

These were the early days of aviation and she had read about
women like Amy Johnson, Amelia Earhart and Beryl Markham who
were pioneers on a par with men, flying long distances
in small bi-planes with open cockpits.
She kept newspaper articles about them, their daring
exploits, defying convention.

Free spirits, the newspapers called them.

Some Women

Some women marry houses
                    Anne Sexton

A place for everything and everything
in its place.
Not a pot or pan misplaced in the spotless
kitchen. No messy piles of newspapers.
No clothes left lying about.
Carpets brushed and clean. Cushions plumped.
Tiles gleaming.
Brass ornaments glittering with polish.
Books banished—they collect dust.
Housebound, they embraced houses.
Before I was married, they would say,
I used to...

Dignity

If my father or father-in-law for that matter
had received an expensive solid gold watch—
an unlikely circumstance—he would have put it
in a box, kept it in a drawer.
His plain Rolex with the leather strap had no
purpose other than to tell the time.

Now it is a different world and how
the old gentlemen in the café sport
large gold watches
on their wrists, some with tiny diamonds embedded
in them. Gold watches that seem to confer some
dignity as if to say, forget the watery eyes, the
unsteady hands, we are not done yet.
Don't disregard us as merely old.

Understanding

I wish, she said, I could believe—the way our grandparents did—
it must be a comfort.
I suppose that's what religion is, she said, comforting.
Something to hold on to, to steady you.

We were waiting in the hallway of the house where the priest lived.
A priest we had never met.
And we were there to arrange a funeral—disturbed by grief
in a difficult, confusing time.

The priest was kind and understanding, his black vestments
a strong symbol of faith when it was most needed.
He had dealt with this situation before, people who needed him
though they had never been to his church—or any other, never or not
perhaps since childhood.

I wish I could believe, they said.

Only Then

In time those clothes in the wardrobe will be
      just clothes,
not her favourite dresses or the suit that cost
      more than they could afford,
nor any other warm remembered things
      just clothes—
to be put in a pile for collection by the Red Cross
or Salvation Army.
      Only then can it be done.

In time the beautiful china cup with a chipped edge
      kept because it was Grandma's tea cup,
will be simply a cracked cup
      because no one will know who loved it
and kept it for so long.
      Only then will it be thrown away.

The Autocrat

Nay, 'am not 'avin that, he would say
whenever his views
        were challenged.
He knew what was right and proper and his
certainty
        was a solid wall of defence
against change.
There was no argument. That was it.

The Immigrant

Someone asks the sausage vendor in the market
where she comes from.
Poland, she says, I miss it.

She misses the home-cured pork, dark
moist rye bread, cabbage and borscht.
She has this other world in her head.

It always comes down to food.

At the Boot Inn—1956

Mr. Morris wore heavy brown boots with metal studs
That scraped on the tiles when
He came into the kitchen.

A farmer, he always had dirt under his finger nails
He would sit by the coal-burning iron range
Eating bread and milk
He had no teeth.

Mustn't grumble, he'd say.
He liked a pint of Guinness from the bar.

Just to Say (Warm Version)

I have eaten a peach from your tree
in Frascati, and the sweetness of summer
juice is running down my arms.
Nothing but the sheer pleasure
of a ripe yellow peach,
till I have devoured it down to the brown
corrugated stone in the centre,
sucking
every last bit of flesh from it.

Forgive me—it was delicious.
sun-ripened in the orchard.

Never

I have never seen the church, only a photograph
carefully composed to hide the fact
that it was a Catholic church not a proper
Church of England.
We all conspired in this—not small—deceit.
In fact it was a huge deception, going far beyond
the matter of the building.
We could rationalise it—had to—because how else
to explain?

How else to explain to ourselves? The guilt lasted
a lifetime, such consolation as there was being
that Mother had not been hurt when so few months
were left of her life.
Much later when the priest spoke gently and they
shared a glass of sherry, she found conversion was
impossible with Mother's feelings in the way.
She could never become a Catholic.

Economy

Some people like things spare, prefer deserts
to jungles, love bare landscapes,
shapes of dunes, their elegant curves.
Love stony hills without vegetation, just
heaps of granite rocks.
They like deciduous trees best in winter,
stark outlines against sky or snow.
A fish skeleton is more interesting to them
than the fish, a snail preferred to a squirrel.
Drawings that leave out all but essential
lines.

This Cold Morning

The front parlour was the coldest room
in the old house.
With its large mahogany sideboard, adorned
by ornate brass candlesticks,
upholstered grey velour sofa and small oak
table, covered with a lace cloth,
it had a formal air,
was seldom used.

This morning, the door to the front parlour
is closed.
People are coming and going, speaking in
hushed whispers.
The children have been sent away, uneasy and
slightly afraid.

Waiting or Walking or On a Crowded Bus

Waiting is part of your life if you are poor or
have trouble making ends meet.
Waiting for buses that never seem to come on time
and are often full when they arrive.
Waiting or walking, whatever the weather, or on a
crowded bus—that's how many people live.

We were in a fancy car in China, in Beijing, lords
of the new road, shared with rickshaws, and bicycles
burdened with all kinds of goods, once a large steel
girder balanced on the handlebars, sometimes with
people being transported on makeshift carriers—
an old lady in a wooden box fastened on the back.
Buses drew alongside, tired faces pressed against the
windows, crowded together, staring at us foreigners
comfortable in our spacious seats.
If you have always ridden in a car you might not
notice—those who are still waiting, walking or on
a crowded bus.

The Shrimpers

There was a stile—weathered to the look of driftwood
        where I used to sit for hours
watching and listening to the sea, loving the open air,
        the open sky and the wide open span of the sea marsh.
And the smell—always the smell of—of fish, salt, seaweed,
        sea pinks, sour and pungent.
Myriads of marsh sounds—sea gulls squawking, cows coming
        to graze, and the rumble and splash of the heavy
horse-drawn carts as the shrimpers returned from the sea
        cutting deeper into the ruts that scarred the mud
from their many passages back and forth over the years.
        The men, exhausted, leaning back against the thick
rope nets slung across the shafts of the carts.
        Glad of the catch, which their wives will boil
ready for sale.
        They lumber up the grassy slope of the seabank then
down to the cinder tracks that lead to the road.

There you will see them at the end of their long day, hear
        the rumble of the carts, the clip-clop of the horses hooves
and the soft thuds of steaming manure dropped on the tarmac
        claimed in shovel-fulls for garden fertilizer.

Family

Eileen and Len in Tanzania, 1967

Len and Eileen pregnant with Roger, 1968

Eileen and Len with children David, John and Sara, and Roger in waiting, 1968

Eileen and Len's wedding, Oct 8 1966

Eileen and Len on their wedding day, October 8, 1966

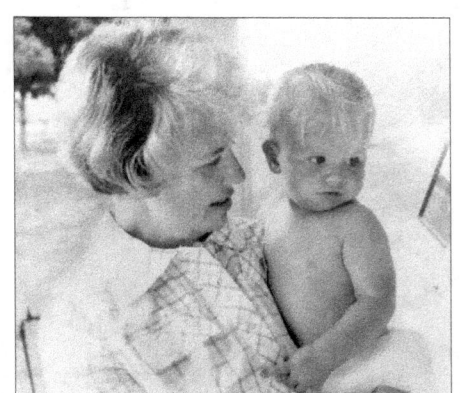

In Tanzania with Roger, 1969

With first husband Arthur, 1951

With her children David, John and Sara, 1959

With daughter Sara and father George, 1965

Eileen with her mother Edith, 1930

Eileen and younger sister Margaret, 1931

Eileen and sister Margaret with their mother, 1932

Eileen and her mother Edith, 1936

Eileen conducting an orchaestra of girls on the wall, 1936

Places

Garden Cottage, East Sussex, England

The house at Longacre, Southport, England

Not on the Map

I found a recent map of Salford with Regent Road
clearly marked—but as part of a new motorway,
not the old road I remember where crowded trams
went clanging on their rails, the air inside
choking with cheap cigarette smoke and seldom
washed clothing. The signs that said "No Spitting"
It was an older world and baths not easy to take,
tin tubs by the fire, money short.

Regent Road—the very name makes me smell the soot
and stale beer from corner pubs, pipe and cigarette
smoke. A few old women still wore heavy black shawls,
hand crocheted, had clogs instead of shoes, as their
mothers did. Even the dogs looked poor, black and
skinny like those in a Lowry painting.

There were some good smells—of yeasty fresh bread
from the bakeries, meat and potato pies, where glass
shelves held sponge cakes filled with jam and cream.
The smell of chocolate and homemade toffee from the
sweetshops whose windows displayed big glass jars of
boiled sweets, pear drops, lemon slices and mintoes.
They have all gone—including the greengrocers where
dirt clung to lettuces, potatoes, carrots, turnips,
fresh from the farms.

There are no more butchers' shops where meat carcasses
hung from the ceiling on large metal hooks, chickens
unplucked, full of feathers and skinned rabbits were
laid out with trays of mutton and beef.
Madam Amy's dress shop, not really French, is not there
anymore. I can find no trace of the Methodist Chapel
where pale green institutional paint on the walls and
pipe organ induced a sense of gloom on Sunday evenings
that made misery seem an integral part of religion.

Cross Lane is still on the map but in smaller letters
and I can't make out if it intersects with Regent Road
as it did when the Labour Exchange was at that corner.

We used to see lines of dejected, down-at-heel young
men in shabby dark workclothes waiting outside, always
waiting—the unemployed in the depression of the 1930's
To our "whys" there was no answer.
So much that is not on the map.

St. Cuthberts' Church

Autumn and the air is still for a waiting moment
Smoke rises from burning leaves
on the other side of the churchyard wall where
the gardener rakes and throws more on the fire

In the old churchyard the headstones on the grass,
cut from the same local sandstone as the church
are mossy with age—a green spongy wetness
soft to the touch

I stand on a pathway of uneven broken stone, injured
stone in the words of one poet, and everywhere is
quiet and still except for the crackling of burning
leaves
A funeral is arranged for this afternoon but for now,
I am alone

Ancient Right of Way

Where once was a path through the woods
You would never know that now
Where all is nettles and brambles
Though some will still remember

You would never know that now
There is nothing to see on the ground
Though some will still remember
Some may notice the stone

There is nothing to see on the ground
No vestige of footprints remains
Some may notice the stone
Half-hidden in the roadside verge

Only the small rough stone
Where all is nettles and brambles
Half-hidden in the roadside verge
Where once was a path through the woods.

Only the Wind There Once

A tarmac road winds around the site of the great
Pyramids of Gizeh—crowded with cars, buses,
donkeys and horses; camels swaying under their
tourist burdens and vendors touting souvenirs of
Horus and Anubis, the old gods.
Black basalt statues, glass replicas of pyramids
and obelisks.
The whole tourist scene—as if the pyramids had
been built just for that.

The camels, look down their noses at this modern
commotion. Haughty beasts of long lineage, their
eyes are patiently wise in the ways of the desert.

The road, with equal disdain, and having nowhere
to go, turns back to Cairo, its mosques and minarets
visible beyond the palms in the sandy fume-filled
haze that blurs the valley.

Once there was only the wind on the plateau, nothing
but sand, thick loose sand, around the pyramids.
Sand that filled our shoes, slowed us down.

Helvellyn

Breathing more easily now after the climb and terror
of Swirral Edge—sheer drop on either side—
we are at the summit where Helvellyn reaches into the
sky—high as gulls lofting in air

More mountains in the distance, a haze of blues and greens

Thousands of feet down, large lakes spanning the flat
valley floors, waters clear and cold as steel

Above us—nothing

We are on the cusp where earth meets sky—meets space

Before

Before the rain, before the thunder,
black clouds on the horizon,
strange light from the underbelly
of the storm
makes the trunks of two royal palms
suddenly look pale, white herons
flying over the lake look whitest white
against the darkening sky,
The trees are still,
immovable as houses.
Before anything happens.
Before the bolt of lightening strikes
from cloud to ground.

Larch Street and the Recent Past

There was this street in the old sepia photograph,
tree-lined, summery,
Its Victorian houses complacent with their bow windows,
solid doors that I remember had stained glass panels
to let light into dark hallways.
Substantial, built to last forever.

And, about half-way down, something new and modern.
A small black, square-shaped motor car—almost
certainly a Ford—parked at the kerb.
A whole street and only one car.

English Summer

Yellow the sunshine on the fields
Buttercups in the grass
It's warm and the days are long

Days when we loved the sun
Never too fierce or strong
Just warm and the days are long

Windows are open, new mown grass
Flowers adorn the streets
Earth feels warm and the days are long

The sun sets late leaving yellow-pink
Skies, cricketers leaving the field
Still warm and the days are long

Happened

In this place where pale stone-washed light
left us sharing space with the stark shapes
of granite hills, mounded dunes of thick,
soft sand, rivers of stones and a single thorn
tree standing up clear of the rugged African
desert, we wait for the jeep to return and
while away the hours breaking up bits of dry
bread to feed to dung beetles who took it
unconditionally, indifferently, as if it had
come to them from above, like manna.

Ecology

Two bright green chameleons were clinging
to the shiny peeled-green upper trunk
of a royal palm,
and I saw they were mating—one scaly body
atop the other,
quiet and still;
their greenness at one with the greenness
of the tree and its huge fronds,
slowly stirring in a light breeze.
It was as if green were an element,
something to be in—like water or mud.

Eclogue

Some people seem rooted in a particular place, so
if you love fir trees
        and the dense dark woods of Maine, ancient
European forests of earthbound mystery and legend, haunting
        smell of pine, knowledge of snow, some
sense of primitive wildewood,
        you are almost certain to disdain the palm,

brash extrovert of trees, blatantly posing on the shore
or simply ornamental,
        standing in streets and gardens, demanding attention,
not content just to exist.

Last Train

It is mid-twentieth century on a small near-deserted
      railway station and the waiting room
seems isolated in the dark,
      like one of those rooms in a large building
that is the last to turn out the lights.

Waiting for the last train late at night a few tired
      souls sit silently on two rows of seats
facing each other, too tired even to stare.
      A naked light bulb hangs from the ceiling
giving out an unnaturally bright light.
      There are workers from the late shift, mothers
with babies and small children lolling against them,
      old men nodding off.

At the borders of sleep that rolls over them
      like a warm blanket,
they are jerked awake at the sound of the incoming
      train,

Still a long way from home.

Word Hoard

The days are drawing out, the old people would say
feeling good that summer was on its way.
The days are drawing out—becoming longer as the
spring equinox approached, often not warm enough,
probably raining, hope deferred, summer some months
away. Drawing out—an odd phrase.

I used to think it had something to do with drawing
the curtains—the swish of it as Father drew the two
sides apart or pulled them together.

When in the autumn the days began to get shorter, it
was said that the nights were closing in and this image,
the swish of the curtains soon to shut out the dark
rose up again.

The words are so much a part of that time and place
they conjure it up after many years have gone by.
Words have such power—carry much more than meaning.
Carrying whole worlds within themselves.

Orange Sun Sets on Ottoman Empire

Have you ever seen an orange peeled methodically
        by long white fingers,
delicate, reticulated, knuckles arched,
        giving the impression of a large white
bone-shelled crab, carefully lifting its claws
        preparing its prey?

It was for me, oddly, part of a discourse on the
        disintegration of the Ottoman Empire, with
dis-in-te-gra-tion very deliberately pronounced,
        each syllable enunciated separately, each
enunciated syllable split off into segments,
        like the orange she was peeling, while points
of the argument were detached, picked apart.

Under the heading: History of Turkey, I wrote:
        Ottomans, then, Oranges.

Street Scenes

When she drew a chair up to the window, she thought of all
those women in black clothes sitting outside
      their hot kitchens, on old wooden chairs in Italian
villages like Positano or Portofino,
      neither inside nor outside but resting in-between
watching the world go by—that necessary sustenance,
      to be in the world, to be with other people.

When it gets warmer she will sit outside too and watch
      the man with the large black dog take his afternoon walk,
see the cars slowing down as they reach their driveways,
      notice the delivery truck making its stops
and people on bicycles enjoying the sunshine.

It is not quite the same as being in Italy, more individual,
      just the odd person sunbathing on a patio,
someone tidying a few flower beds or watering the lawn.
      Otherwise people go by in their cars, not waving
or even seeing. After awhile neighbours or people walking
      their dogs get used to noticing and regular greetings
become part of the day, vestiges of those ways of being human
      in Italy where old men sit outside playing dominoes
and old women sit outside on kitchen chairs—
      ways of being alone and together at the same time.

Change

With father George Hadley, 1932

Isolation

Only the broken buildings of the ISOLATION HOSPITAL
        remain now—shards of shattered glass,
bricks and wooden beams in the grass;
        through an empty window frame we see
an iron bedstead left in a corner.

Sad reminder of the children sick with diphtheria
        and scarlet fever who were put here
in quarantine.

We came upon it when cycling along dirt tracks that
        cross the peat moss outside the town.
It stood on a patch of higher ground surrounded
        by trees.

The name spoke to that earlier time when there was
        a grim sounding lunatic asylum,
and a home for fallen women where girls did laundry
        work on the cheap and were allowed to keep
their babies.

Transience

It was sad to see Longacre without the farm and fields,
    Longacre itself the name of a field,
now nothing but another suburban street—all that glorious
countryside gone.
    Where once there were hedges of prickly brambles
and wild roses—the ones with flat white single petals—
fields of wild flowers, narrow winding lanes where small
cottages, some with thatched roofs, seemed to nestle
alongside.
    Now, there are impersonal brick buildings and the old
farmhouse—renovated and unreal—stands like a dowager, all
dressed up with nowhere to go.
    Outside is a FOR SALE notice.
Apartment buildings have replaced the old cowsheds and barns
where hay was stored.
All that wild and lovely landscape was once—when we were young—
the settled and certain background to self.
    We thought it would never change.

Wildness

> *What would the world be, once bereft*
> *of wet and of wildness? Let them be*
> *left, O let them be left*
>   Gerard Manley Hopkins

We ran down the lanes to the sea—Knobbals, Threlfalls, Petes,
      Cockle Dick's Lane holding our noses at the stink
of manure from pigsties on the farms, and played on the shore
      where surging waves wet the brown sands,
and on the sour fish-smelling marsh pink with sea thrift
      musty as honey.

We picked wildflowers in the fields, scratched our arms on
      blackberry brambles, looking for wild fruit
in the hedges.

We were ghosts in the making, careless of time
      that would take away fields and hedges, lanes
and farms and the old earthbound slate-roofed houses,
      tame the marsh into a bird sanctuary,
leaving the seashore cut off by a new road, waves pounding
      the sands as always.

Eileen Berry

October Comes Round Again

You come to me in broad daylight,
there at the dining table,
eating a large Spanish onion with
pepper and melted butter,
while the good smell of rising
dough comes from the bowl, covered
with a clean kitchen towel, that
you put in front of the fire.
Soon we will go for a walk on the
marsh, be blown about by the wind.

Waiting for Something Terrible to Happen

The day seemed like Sunday—boring, but it was not
Sunday. No Wesleyan Chapel, singing about Jesus.
Only the smell of Sunlight Soap on newly washed hair.

A commotion of voices, heavy dark overcoats were
shifting against the front door, black polished shoes
moving over the red Turkey carpet in the hall.
Adult conversations whispered.
The visitors gave me a large ceramic doll which had
belonged to Eva.

I hated its hard pink face, its stiff cold feel and
coarse hair; most of all I hated its glass eyes that
opened and shut when it was moved up and down.
Held on my knee, it suddenly fell to the floor and the
dreadful eyes fell back into its head, rattling like
marbles and leaving empty holes for eye sockets.
Silence followed.

Waiting for something terrible to happen, I felt as if
I had been left—at someone else's house
with people I didn't know.
I was homesick for a different self—
for a different day.

Chance

Three small petal-like leaves, pale magenta, almost
translucent, floating on the water.

Two lie flat on the surface; one, dry, stands up,
a tiny sail, so fragile—butterfly of a boat.

Leaves from the bougainvillea blown on the pool,
one little spray, alighting alone—mystery.

Only one—its perfection more apparent—
as in a Japanese flower arrangement, minimalist.

Like a sudden glimpse—of deer or a snowy owl—
and we the lucky ones.

Singular

The jacaranda tree is spreading purple
        all over itself, all over
its delicate silver-grey branches, all
        over the street.
Purple leaves fall in heaps, bushels of them,
        shovels-full.

Leaves to shuffle through, to rustle
        like fallen leaves of autumn
in the north.

The jacaranda tree is singular,
        with none of the myriad greens
of Florida.
        Overwhelmed by riotous royal purple,
it stands apart,
        insisting on its purple self.

Reflections

The setting sun drops into the lake
like a ball of fire,
spreading a comet-like trail of light
till, by contrast,
the water becomes black, opaque, changed,
unknowable

the way the sea at night is changed
to an impenetrable blackness,
as it retreats into the sky, losing its
horizon.

Reality

A small tuft of dried grass and I loved
its straw-like texture, its spare shape.
It resembled an elongated grasshopper,
its thin angled limbs dominated by its
large poised head,
Had Giacometti worked with straw, he might
have created a form such as this,
I placed it carefully on the mahogany desk
in the hall.

One day someone saw a small dried up tuft
of grass, threw it away.

Drama

Unlike human drama, this is raw and unedited,
unscripted, the ending not even half-known.
Grey November skies stormy with passion, in
constant motion, changing panoramas.
The natural world where we are unsure of our
parts, learning as we go.

Hiatus

White morning, the whole place blanketed in powdery
new-fallen snow, soft and light as goosedown.
No horizon.
No boundaries visible between fields
and roads.
Everything buried under pillowing mounds
of fresh snow—only shapes of cars, walls
steps emerging
No sound but that of cracking branches,
sudden swish of dislodged snow hitting the ground.
Stillness pristine.
Before the snowballs, footfalls, shovels and ploughs.
Before the digging out.
Standstill

Persistence

I admire the purple foxglove—how it thrives
in the azalea bed,
as if it had every right to be there, thrusting
itself up again and again,
no matter how often it is rooted out, thrown
on the compost heap.

It is quite beautiful—with its purple-white
tiny bell-shaped flowers on tall furred stems,
but unwanted in this place,
A weed—wild thing—sown by the wind that
disperses where it will,
paying no heed to boundaries.

The Red Poem

This horrible but superb painting
the parable of the blind
without a red
in the composition...
              William Carlos Williams

Is there a poetry of red heard by the blind?
Percussion of red, resonant, redolent,
potent red,
that marks a red shift in the stars, feels
lichened touch on stone, litmus of the bone,
alizarin's taste on the tongue?

Deep, deep red, red that feeds all the senses,
red that is all of redness.
Cardinal, sensuous Roman red, ancient
velvet red-dark scent of roses.
Drumbeat sounds of red, red pepper
burning the throat.
Red that baits the bull.
Lucky, joyful red, slithering, soft, Chinese
wedding silks, exhilarating blood-red wine.
Treacherous red of poppies and poison berries.
Crimson, carmine, cerise, cayenne,
safflower, scarlet, vermillion.
All the dangerous warning, compelling, ripening,
happy, fortunate, words for red.
Red thoughts becoming red words in language
Prometheus stole with fire from the gods.

Rain Dance

Wild, cold, coursing rain, pounding leaves
        and flowers, seemingly intent
on damage.
        Pouring down gutters, turning lawns
into marshland, making rivers of roads.
        Rain beating on windows, thrumming,
thrumming, thrumming.
        Cold, drenching, miserable rain, wetting
clothes, filling shoes till they squelched,
        soaking people to the skin.

Absence

There is a photograph on the wall, of an empty
seat in a garden, and a painting of a petrol
station, isolated, abandoned in a desert
landscape—both suggesting absence—but when
her child has just left home for the first time
to live away, she stands in the empty bedroom,
absence is almost presence.
Love hurts and she is lame, limping inside.

Separation

That woman feeding her baby could be me.
I almost reach for lanolin in the lumpy
bag I used to carry.

Milk spurts as she shifts the child,
protecting her swollen breasts with the soft
baby blanket,
and I reclaim an old discomfort from that
milk-safe time.
Physical closeness—before the growing up,
the moving away.

Arts Poetica

One shoe came off so I was at a disadvantage and
I moved in closer but couldn't hear what the man
said. He was eating an apple, not the ordinary kind
because there was green sticky stuff round his mouth
and he thought I was deaf and spoke more loudly and
I still didn't get it but it sounded like you aren't
some kind of health nut are you and I couldn't find
my glasses so when someone introduced me to my
daughter I thought she was someone else and suddenly
there was a woman giving a poetry reading who
was important or at least I think so, everyone was
listening to her and I drifted off to where they
were selling T-shirts on the edge of a steep set of
steps which I didn't want to go down because they were
intentionally imperfect like an Arabian rug and had
broken stones in places. You might think this is
a dream or perhaps a poem. If it's a poem I want to
change it, but I'm stuck. The words keep dropping off
the page and I'm missing a black patent leather left
shoe

Notes

"The Red Poem" quotation is from William Carlos Williams' poem "Pictures from Brueghel"—*The Parable of the Blind*

"Just to Say" is also the title of a poem by William Carlos Williams about eating plums cold from the icebox. Hence the qualification here "warm version."

"Ancient Right of Way" refers to a customary right of access for the public in England and Wales mainly to walk and cycle in the countryside, by paths, tracks or lanes.

"Helvellyn" refers to a peak over 3000 feet high in the English Lake District.

About the Author

Eileen at Garden Cottage, East Sussex, England, 2004

January 8, 1927 - March 19, 2016

Eileen Berry is a geographer with a first degree from University College, London and a PhD from Clark University in Massachusetts. Geography has a strong presence in her poetry. She grew up in England and also lived many years in Africa. These places are at the heart of this book. Her poems attempt to bring back the sense of places that are now lost or are fast disappearing along with peoples' memories of them.

Some fifty of her poems have been published in journals such as *The Notre Dame Review*, *Rattle*, *The Liberty Review*, *Tar River Poetry*, *Bayou*, *Lynx Eye*, *Eclipse* and *The Louisville Review*. She has been accepted twice as an Associate at The Atlantic Center for the Arts, once in 1992 when Amy Clampitt was the poet-in-residence and again in 1997 with David Lehman. She was nominated for the Pushcart Prize for the second time in 2005.

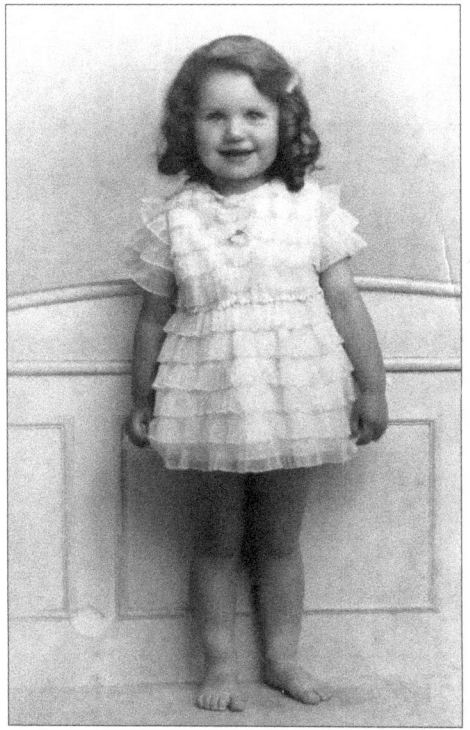

Eileen, circa 1930

Eileen, 1933

Wedding day with Arthur, 1950

Eileen, 1945

Eileen, 2015

www.ingramcontent.com/pod-product-compliance
Lightning Source LLC
Chambersburg PA
CBHW050043080526
44586CB00014B/1431